GARETH STEVENS
VITAL SCIENCE
Physical Science

SCIENCE AND SOCIETY

by Robert Snedden
Science curriculum consultant: Suzy Gazlay, M.A., science curriculum resource teacher

Please visit our Web site at: www.garethstevens.com
For a free color catalog describing Gareth Stevens Publishing's list of high-quality books and multimedia programs, call 1-800-542-2595 (USA) or 1-800-387-3178 (Canada).
Gareth Stevens Publishing's fax: (414) 332-3567.

Library of Congress Cataloging-in-Publication Data

Snedden, Robert.
Science and society / Robert Snedden.
p. cm. — (Gareth Stevens vital science. Physical science)
Includes bibliographical references and index.
ISBN-13: 978-0-8368-8089-2 (lib. bdg.)
ISBN-13: 978-0-8368-8098-4 (softcover)
1. Science—Social aspects. 2. Scientists—Social aspects.
3. Creative ability in science. I. Title.
Q175.5.S586 2007
303.48'3—dc22 2006034461

This edition first published in 2007 by
Gareth Stevens Publishing
A Member of the WRC Media Family of Companies
330 West Olive Street, Suite 100
Milwaukee, WI 53212 USA

Produced by Discovery Books
Editors: Rebecca Hunter, Amy Bauman
Designer: Clare Nicholas
Photo researcher: Rachel Tisdale

Gareth Stevens editorial direction: Mark Sachner
Gareth Stevens editors: Carol Ryback and Leifa Butrick
Gareth Stevens art direction:Tammy West
Gareth Stevens graphic design: Scott Krall
Gareth Stevens production: Jessica Yanke and Robert Kraus

Illustrations by Stefan Chabluk
Photo credits: Apple: p. 4; Cellnet: p. 27; CFW Images: pp. 35 (Chris Fairclough), 41(Paul Humphrey); CORBIS: pp. 32 (Roger Ressmeyer), 33; Department of Defense: p. 42 (Jon Gesch, USN); ESA: p. 25 (NASA); FLPA: p. 10 (Mitsuhiko Imamori); Getty Images: Cover (Boris Lyubner), pp. 5 (David Trood), 6 (Stewart Cohen), 7 (Jae Rew), 13 (Dr Hans Gelderblom), 15 (Frederick M. Brown), 16 (Topical Press Agency), 28 (Hulton Archive), 37 (Chris Wilkins/AFP), 40 (Toru Yamanaka/AFP); Honda UK: p. 34; Istockphoto: pp. 8 (Steve Beckle), 12 (Paolo Florendo), 17 (Adam Korzekwa), 19 (Andrei Tchernov), 20 (Dominik Pabis), 21 (Steve Grewell), 26 (Lance Bellers), 29, 30 (Eliza Snow), 31 (Jason Stitt), 43 (Gina Smith); Philips: p. 23 & title page; Photodisc: p. 39; Science Photo Library: pp. 18 (Dr Yorgos Nikas), 22 (Sovereign, ISM).

Printed in Canada

1 2 3 4 5 6 7 8 9 10 10 09 08 07 06

TABLE OF CONTENTS

Words that appear in the glossary are printed in **boldface** type the first time they appear in the text.

Cover: This illustration by Boris Lyubner shows the application of science and technology to a variety of fields, including business, medicine, engineering, and research.

Title page: An ant carries a microchip in its jaws, showing just how small high technology can be.

Introduction

Science is unarguably one of the greatest achievements of the human race. Our world would be totally different if it weren't for science. In this book, we're going to see some of the many ways in which our lives and our society have been changed by science. Science is the best way we have of finding things out.

In the modern world, **technology** is driven by science. Without it, we wouldn't have many of the things we take for granted. There would be no cars, no television, no MP3 players, no cell phones, and no digital cameras.

Every year, MP3 players and other gadgets seem to get faster, smaller, and more powerful.

What do we think of when we think of scientists and science? As we look at science and its place in society, perhaps we should keep one question uppermost in our minds. What does society want from science?

One thing we expect in life is progress. We expect science to come up with something more than we had before and expect that what scientists discover today will be used in some fantastic new gadget tomorrow. This year's cell phones will be smaller and have more gadgets than last year's models. Computers will be faster and cheaper. Automobiles will be more fuel efficient.

Science and Uncertainty

Another thing we expect from science is certainty, especially certainty of safety. The science we learn in the classroom seems very definite. We learn equations and formulas with the understanding that they will always be true. We see that if we mix chemical A with chemical B, we will always get the same reaction. As a result, people may come to expect this certainty from all their dealings with science, even from the cutting edge of discovery.

Science today is about more than what goes on in the **laboratory**. It touches our lives in many ways. We have to know what the limits of science are and accept that sometimes science doesn't have definite answers to give, especially when dealing with big issues, such as **climate change**,

gene therapy, and energy shortages. These things can't be reduced to simple equations.

Science has a duty to keep us informed about the way it goes about its work. We, as citizens, need to be aware of what science is doing. We should respect the fact that within their fields of expertise scientists may have more knowledge than the rest of us. But this doesn't mean that scientists are people who cannot be challenged.

Do doors exist that should forever remain closed to science? Science is about gathering facts and expanding our knowledge of the world and our place in it. One thing science won't do is to decide for us whether a thing is good or bad. An aircraft, for example, can be used to take us on vacation, or it can be used as devastating terrorist weapon. The discovery of **antibiotics** is another example. Antibiotics have saved the lives of many people, but their overuse has led to the rise of seemingly unstoppable bacteria, called superbugs. Would anyone seriously suggest that we should never have had aircraft and antibiotics? As a society, we have to decide what to do with the knowledge that science has given us already and will give us in the future.

ILLUMINATION

"Science knows no country, because knowledge belongs to humanity and is the torch which illuminates the world."

— Louis Pasteur (1822-1895)
French microbiologist

Not all science is carried out in the laboratory. It can even take you to the slopes of an active volcano.

CHAPTER 1

Science and Scientists

What do you think of when you think of scientists? Do you picture mainly men, generally in white coats, doing weird and incomprehensible things in a laboratory? It might come as a surprise to you, but scientists look just like everyone else. If you saw a scientist in the shopping mall, you'd never know! Scientists can be men or women, young or old, and of any ethnicity or background.

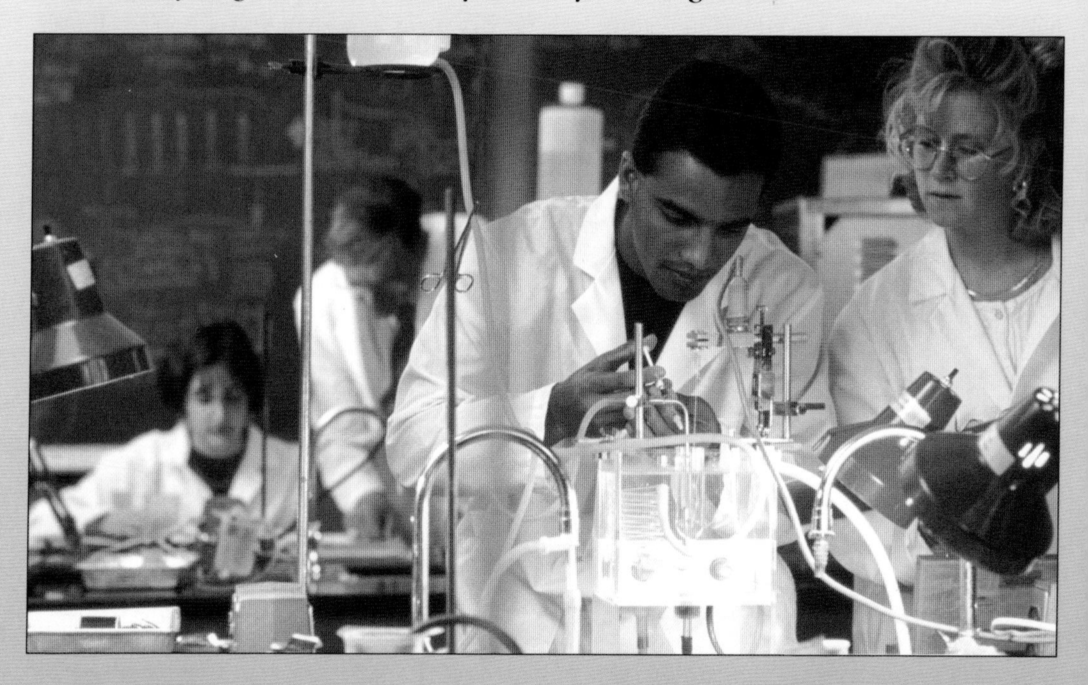

These students are learning the skills they will need to take science forward in the future.

The Practice of Science

Scientists can be found at work in many different places. They are not always in the laboratory. There are scientists in offices, in factories, and on farms, at crime scenes, and on wildlife reserves. Science goes on everywhere.

So what makes someone a scientist? Let's look at what scientists actually do. Scientists believe that the world works according to a set of rules that we can discover and understand. Scientists gather information through careful observation. They look at what's happening in the world around them and try to make some sense out if it.

You've noticed, for example, that if you throw a ball into the air, it always comes back down again. You wonder why that happens. Scientists in the past have wondered just the same thing. Science calls a possible explanation for why something happens a **hypothesis**. If the explanation is correct, then it should be possible to make **predictions** based on that hypothesis. If your hypothesis is that "everything falls to Earth," for example, you might predict that whenever anyone anywhere throws a ball into the air, it will come back down.

Next, you can test your prediction. Scientists do this by setting up **experiments**. An experiment is a way of finding out if your predictions are correct. You might experiment by throwing a baseball, a basketball, a football, and a tennis ball into

If you throw a ball up in the air, it always comes back down. Science can tell us why.

the air to see if the size of the ball makes any difference. You might also see if it makes any difference if you throw the ball really high.

If you find that your hypothesis always holds true, you might go on to propose a theory. A theory is the best explanation that can be found for why something happens. The story goes that Sir Isaac Newton watched an apple fall from a tree and realized that there had to be a force that made the apple head for the ground instead of floating away. It didn't matter how tall the tree was—every apple fell. Newton called this force that pulled things to the ground **gravity**. His genius was in realizing that gravity didn't stop at the treetops. It was a force that reached all the way to the Moon

Gravity, the same force that brings a ball back to the ground, keeps the Moon in orbit around Earth.

and kept it in orbit around the Earth.

From this idea, Newton came up with his **law** of universal gravitation, which states that every object in the universe attracts every other object with a gravitational force. It's the force of gravity that makes your ball come back to Earth.

A law of science is an explanation or a statement that always appears to be true. Science doesn't stand still, however. Newton's law of gravity and his three laws of motion are excellent at explaining why objects move the way they do. However, in 1905, Albert Einstein came along and showed that for objects traveling at speeds approaching that of light, Newton's laws didn't apply any more. Newton wasn't wrong—he just couldn't imagine or foresee the limits of his laws.

THE SCIENTIFIC METHOD

The scientific method is the process scientists follow on the path to new knowledge and discoveries. The steps on the path are:

- Observation/Research: What do you see?
- Hypothesis: Why do you think it happens?
- Prediction: If you're right, what will the outcome be?
- Investigation: What can you do to test your predictions?
- Conclusion: What did you learn from your investigations?

SCIENCE OR POETRY?

"In science one tries to tell people, in such a way as to be understood by everyone, something that no one ever knew before. But in poetry, it's the exact opposite."

— Paul A.M. Dirac (1902-1984) Nobel Prize-winning physicist

ISAAC NEWTON

Isaac Newton (1643–1727) was born in Woolsthorp, England. His father died before he was born, and Newton spent most of his childhood with his grandmother. He attended Cambridge University, but much of his best work was done when he spent two years at home when the university was closed because of the plague. During this period, he studied the spectrum, invented the reflecting telescope and observed the Moon and planets, and developed his laws of motion and gravity.

Newton became a professor at Cambridge in 1669 and was elected president of the Royal Society in 1703. He had many arguments with his fellow scientists and hated to be criticized, but his achievements place him close to the top of the list of the greatest scientists who ever lived.

Darwin's Hypothesis

The British naturalist Charles Darwin (1809–1882) was an observer of the natural world. He was well aware of the relationships between flowers and the insects and birds that pollinated them. All flowers need to be pollinated. Some flowers are able to pollinate themselves, and some can be pollinated by wind or water. Most, however, rely on animals—usually insects or birds—to transfer pollen from one flower to another.

Usually there is a reward for the pollinator in the form of sweet nectar. Darwin discovered a flower on the island of Madagascar that held its nectar at the end of an 11-inch (28-centimeter) long tube. No insect known at the time had a **proboscis** long enough to reach it. Based on the evidence that the flowers were being pollinated somehow, Darwin formed a hypothesis. He suggested that there must be a moth somewhere with a proboscis 11 inches (28 cm) long. He never found the moth to prove his hypothesis, but in 2005, the moth was in fact discovered and Darwin's hypothesis was proved right more than 120 years after his death.

THAT'S FUNNY...

"The most exciting phrase to hear in science, the one that heralds new discoveries, is not Eureka! (I found it!) but rather, 'hmm. . . . that's funny. . . .'"

— Isaac Asimov (1920–1992), science writer

Darwin suggested that a flower with a long nectar tube would need a long-tongued insect to pollinate it, even though he'd never seen such an insect. He was proved to be right.

Tried and Tested

Science is not likely to accept things at their face value. The word science comes from the Latin word, *scientia*, which means "knowledge." Scientists want to know about things. They carry out tests and experiments—sometimes repeating them many, many times. They make very careful measurements using the best equipment available. They carefully record exactly what they did and what the results were. They do this so that others can repeat the experiments. This is one of the most important things about science. Everything can be checked and repeated. If the experiments don't back up the hypothesis, it isn't necessarily a bad thing. We learn by knowing which ideas are wrong as well as by knowing which ideas are right. We can then make adjustments and try again until we come up with a hypothesis that can be supported.

Asking Questions

Scientists are always asking each other questions. When we are dealing with scientists or reading about their work in newspapers and elsewhere, we should be asking the same questions.

- How do you know that to be true? Is it just something you believe, or is it something you've done research and experiments on?

- How were your experiments carried out? Are you sure that your measurements and results are accurate? Are you at all uncertain about any of your results?

- Did everyone else who tried your experiments get the same results?

- Can you be sure that your conclusions are right? Could you possibly look at your findings in a different way?

- Does anyone disagree with your findings, and, if they do, why?

WHAT DOESN'T SCIENCE DO?

Science can tell you why an apple falls and why a baseball goes flying off when you hit it with the bat. It can't tell you why you enjoy the smell of apple pie so much or why it feels so great to hit a home run. Science can explain how vibrating guitar strings send out waves of compressed air that eventually reach your ear, where you hear them as sounds. It can't explain why listening to your favorite song makes you feel good. Science can tell you a lot about the world, but there's more to the world than science can tell us.

Science and Health

One of the many areas of our lives that have been changed beyond measure by advances in science is medicine. Developments in diagnosis and treatment have saved many people's lives and improved the quality of life for countless others. The medicine of today would probably be almost unimaginable to the doctors of a hundred years ago.

Immunization

One of the greatest discoveries in medical science is **immunization**. There's a good chance that you've been immunized against a number of diseases. When you are immunized, you're given a **vaccine**, usually by injection. A vaccine is usually a weakened form of a disease-causing organism. This means that it won't actually make you ill, but when it gets into your body, the body's defense mechanisms spring into action. These mechanisms learn to recognize the disease and how to deal with it. Then, if you ever get exposed to the active form of the disease, you're ready to fight it.

Over the last two centuries, vaccinations have saved countless people from illness and even death.

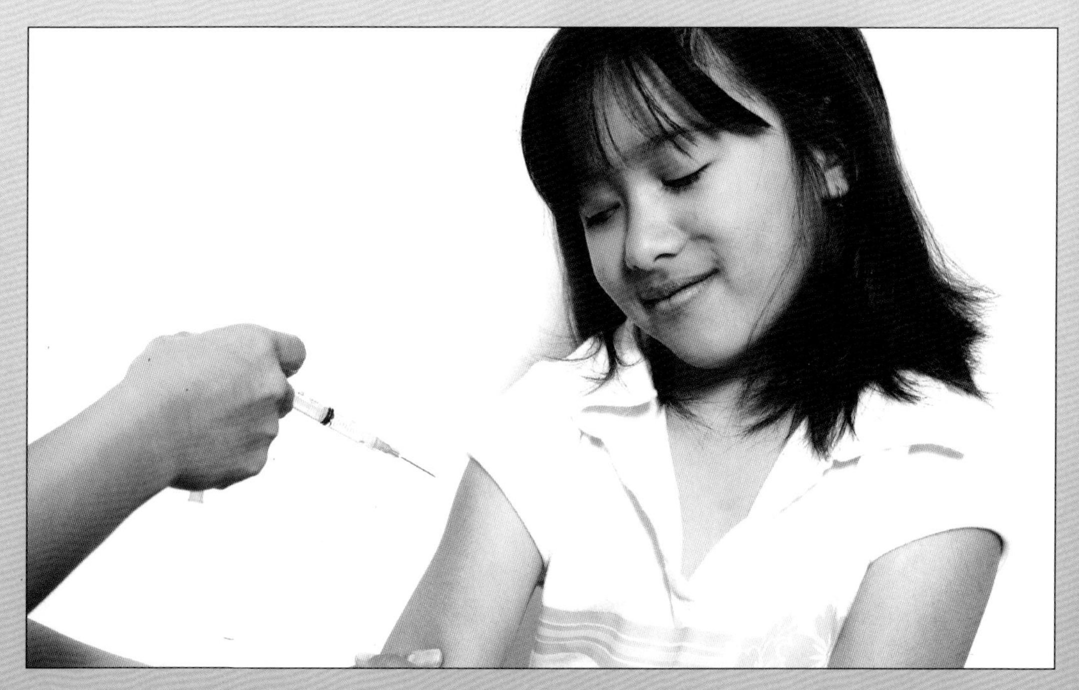

Edward Jenner, an English country doctor, discovered immunization. Like all good scientists, he was observant. He noticed that people who had been infected with a mild disease called cowpox almost never caught the much more serious disease of smallpox. In 1796, he did an experiment to see if the cowpox was giving people protection against the smallpox.

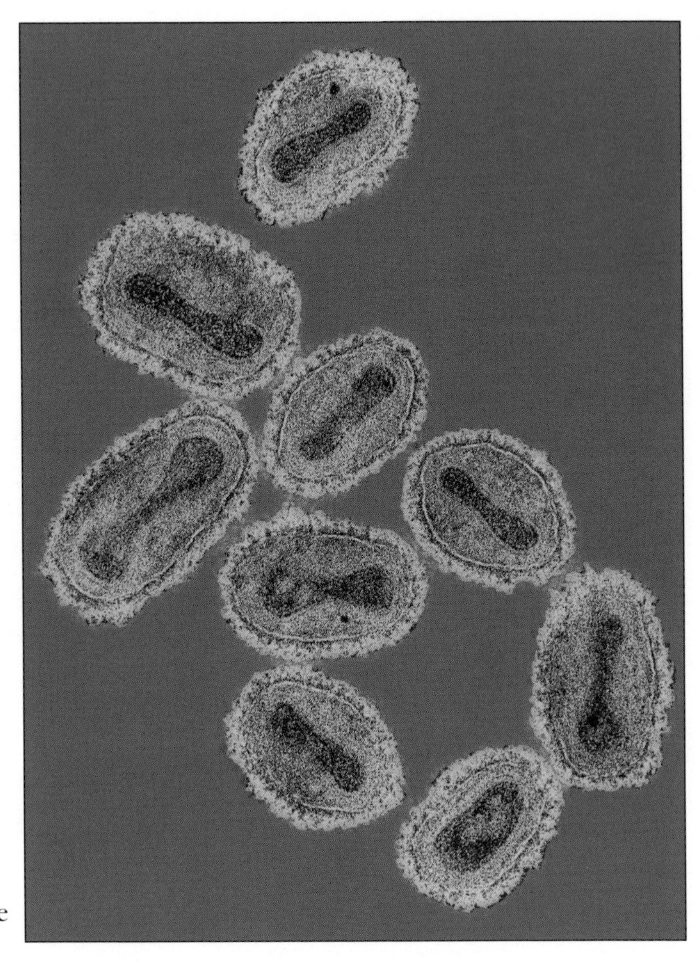

The virus that causes smallpox is seen here magnified 65,000 times.

Jenner took **pus** from the spots of a person suffering from cowpox. Then, he used a needle to scratch the pus into the arm of an eight-year-old boy, James Phipps. James became ill with cowpox, but after a few days he made a full recovery. Jenner then injected James with deadly smallpox. He was unaffected by it. Jenner's discovery would change medicine forever. He coined the word vaccine from the Latin *vacca*, which means cow, and called the process vaccination.

By 1800, more than 100,000 people had been vaccinated against smallpox. The practice of vaccinating had spread across Europe and into America. Benjamin Waterhouse, a doctor with Harvard's new medical school, pioneered vaccination in the United States. President Thomas Jefferson also supported the idea. By 1979, scientists declared that smallpox, a disease that had caused millions of deaths, had been defeated across the world.

Making medicines

Most people use medicines at some time or another, even if it's just taking something for a headache. Others need to take medicine every day to treat conditions such as allergies, asthma, or diabetes. Doctors can prescribe thousands of different kinds of medicine for many different conditions. Where do they all come from?

First, scientists do research to find out what is actually causing a disease. Once they have found out whether the cause is a **virus**, a **gene** not working properly, or something else, they can begin to think about how to deal with the problem. Researchers will look for substances that can prevent, treat, or cure the disease. Sometimes these compounds will already exist in nature. Many medicines have come from plants, for example. Sometimes the researchers look for ways to boost the body's natural defenses by producing a suitable vaccine. They often design a new compound from scratch, however. High-powered computers are vital in this process, and scientists can design new compounds on the screen. Thousands of potential new medicines may be tested and rejected before one goes through to the next stage.

Before a new medicine can come into use, two vital questions have to be answered: Does it work? Is it safe? A great many expensive and time-consuming tests will be carried out to answer these questions. First, the compounds will be tested in the laboratory using **cells** and **tissue cultures**. This involves bringing the compound into contact with small samples of living material. Next, any compounds that still look promising will be tested on animals.

Animal Testing

Animal testing is one of the most controversial issues concerning science and society. We all want to be well and have medicines available when we are ill. Some people believe, however, that testing medicines on animals is wrong. For one thing, they say, it is cruel to the animals. Another argument, however, is that humans are different from other animals. The results of animal tests, they suggest, can't be used to say what will happen if humans take the new medicine.

Both sides have strong arguments. Most animal tests are carried out on mice and rats that spend their whole lives in laboratories and are very well looked after. Modern medicines have helped to cure or prevent diseases that used to kill thousands of people. On the other hand, is it right that animals should suffer at all for our benefit? Shouldn't scientists be concentrating on research that doesn't involve animals? It isn't easy to find effective alternatives to animal testing. Sometimes, it may be the only option scientists have. Researchers are currently looking into the possibility of using computer models to test the effectiveness of new drugs. Many hope that in the future fewer tests will have to be carried out on live animals as a result.

Protestors demonstrate against animal testing at an event sponsored by the Natural Resources Defense Council on May 10, 2002 in Los Angeles, California.

Genetic Medicine

Some of the greatest advances so far in medical care have come through our increasing knowledge of **genetics**. Genetics is about the study of **heredity**, the passing on of characteristics from one generation to the next. It explains why we all have some of the characteristics of our parents. Genetics has a very valuable part to play in health care.

Prince Alexei of Russia (1904–1918) (seen here with his father Tsar Nicolas II) inherited the genetic disease hemophilia, which prevents blood clotting, from his mother.

Science is discovering more and more about genetics. As this happens, the role genetics will play in diagnosing and treating diseases will become increasingly important. The way in which we respond to disease depends on our genes because they determine what our bodies' defenses are. Sometimes, faulty genes are actually the cause of a disease.

Geneticists, the scientists who study genes, are tracking down the tiny errors in genes that may make us vulnerable to illness. More and more tests are becoming available that allow doctors to determine what illnesses a person is most at risk of getting. With this knowledge, you can take steps to avoid coming into contact with these illnesses and reduce possible risk factors, such as diet.

A person's genes also determine how he or she will react to different medicines. Sometimes a medicine that is effective for one person will have no effect on another. Knowing what's right for a particular patient can save time and money and help the person get well faster. Genetic tests can also be used to plan safe treatments. About two million Americans every year become ill because they have a bad reaction to medicine they've been given. So knowing about genes can help to prevent people being given medicines that just don't work or that might actually cause them harm.

Gene Therapy

Gene therapy is a technique with the potential of treating people with genetic illnesses. It involves inserting "good" genes into the patient's cells to take over from those that aren't working properly. So far, progress has been slower than hoped in a field that could hold great promise.

The crucial difficulty to overcome lies in getting the genes safely to the target cells in the patient's body. One way that has been tried is to use a virus to carry the replacement genes, but this can be risky. In 1999, eighteen-year-old Jesse Gelsinger of Tucson, Arizona, died as a result of his body's response to the virus that was used to deliver the genes to his cells. His death was a blow to the advancement of gene therapy.

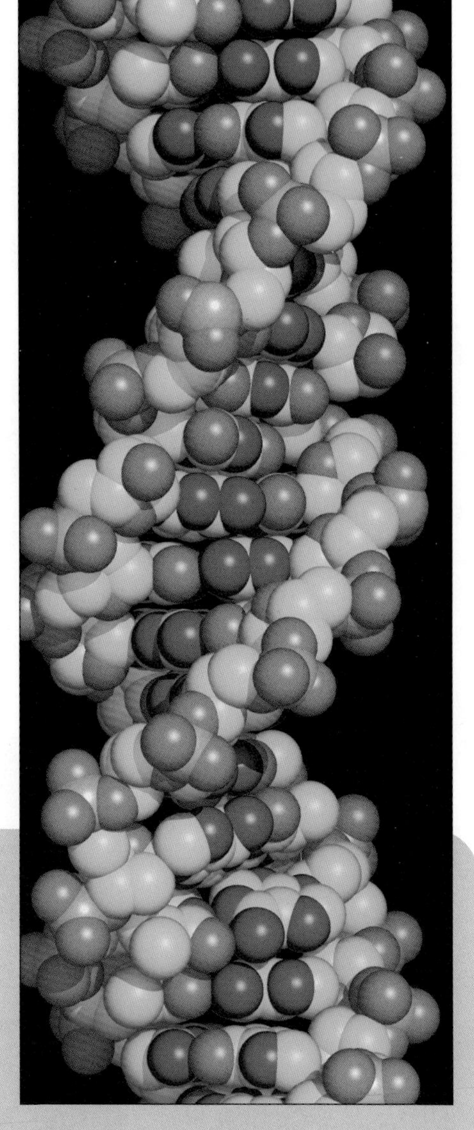

NOT QUITE THE SAME! Take any two people from anywhere on the planet and their DNA will be 99.9 percent the same. It's just that tiny 0.1 percent that makes us different.

Q&A WHAT IS A GENE?

In nearly every cell in your body, there is a giant molecule known as **DNA** (right). DNA is short for deoxyribonucleic acid. DNA carries a set of instructions for making proteins. Proteins are essential parts of your body. They regulate chemical reactions, fight infections, and provide part of your body's structure. The set of instructions for a single protein is a gene. Genes decide if you're male or female. They influence your height, your eye and hair color, and other features that make you, you. When the instructions are clear, everything runs well. Sometimes, however, the instructions are faulty. This can lead to serious, life-threatening conditions such as cystic fibrosis and some forms of anemia.

Another setback occurred in 2003 when French researchers reported that they had successfully cured four boys of "boy in the bubble" syndrome. This name comes from the fact that people with this problem have an ineffective **immune system**. They must be kept isolated from contact with anyone who might expose them to illnesses, diseases, or even slight infections such as a cold sore. Any of these could harm the patients. However, within a few months of the announcement of success, it was discovered that some of the genes had been delivered to the wrong places. Two of the boys developed leukemia.

In the United States, the Food and Drug Administration (FDA) allows research on gene therapy to be carried out. As yet, however, the FDA has not approved any gene therapy products for sale in the country.

In 2006, scientists at the National Institutes of Health in Bethesda, Maryland, made an exciting breakthrough. They successfully treated two patients suffering from a type of cancer by genetically altering the body's killer T cells (part of the immune system) so that they would attack the cancer cells. If gene therapy can be used in treating cancer, it will be a big step forward.

Stem Cells

Your body is made up of a great number of different cells. Some are nerve cells, some are blood cells, some muscle, some skin, and so on. However, you started life as an **embryo**—a collection of cells that were all the same. These embryonic cells are called **stem cells**. They have the potential to become any of the different cell types. As the embryo grows, the stem cells become more specialized. For instance, some become blood stem cells, which can develop into red blood cells or white blood cells. These specialized stem cells are found in the body throughout life. The purpose

A human embryo at a very early stage of development. Each of these cells is a stem cell, with the potential to develop into any other kind of human cell.

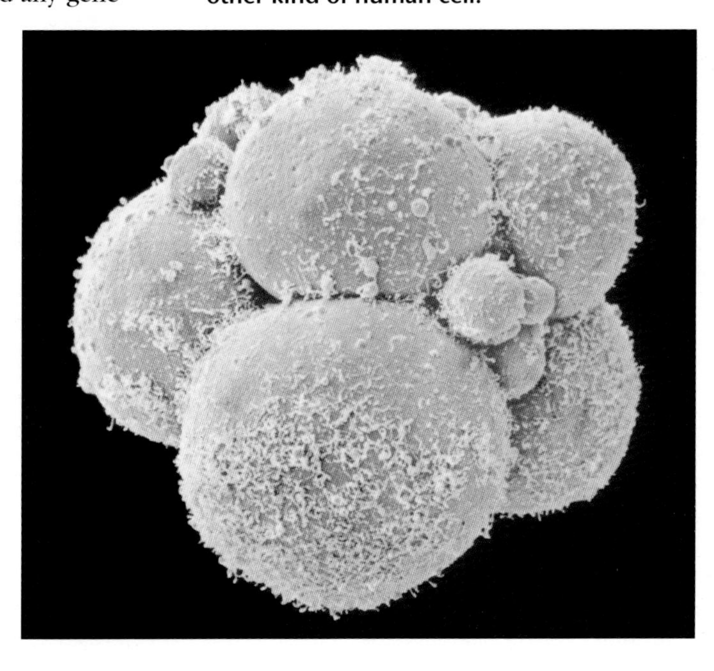

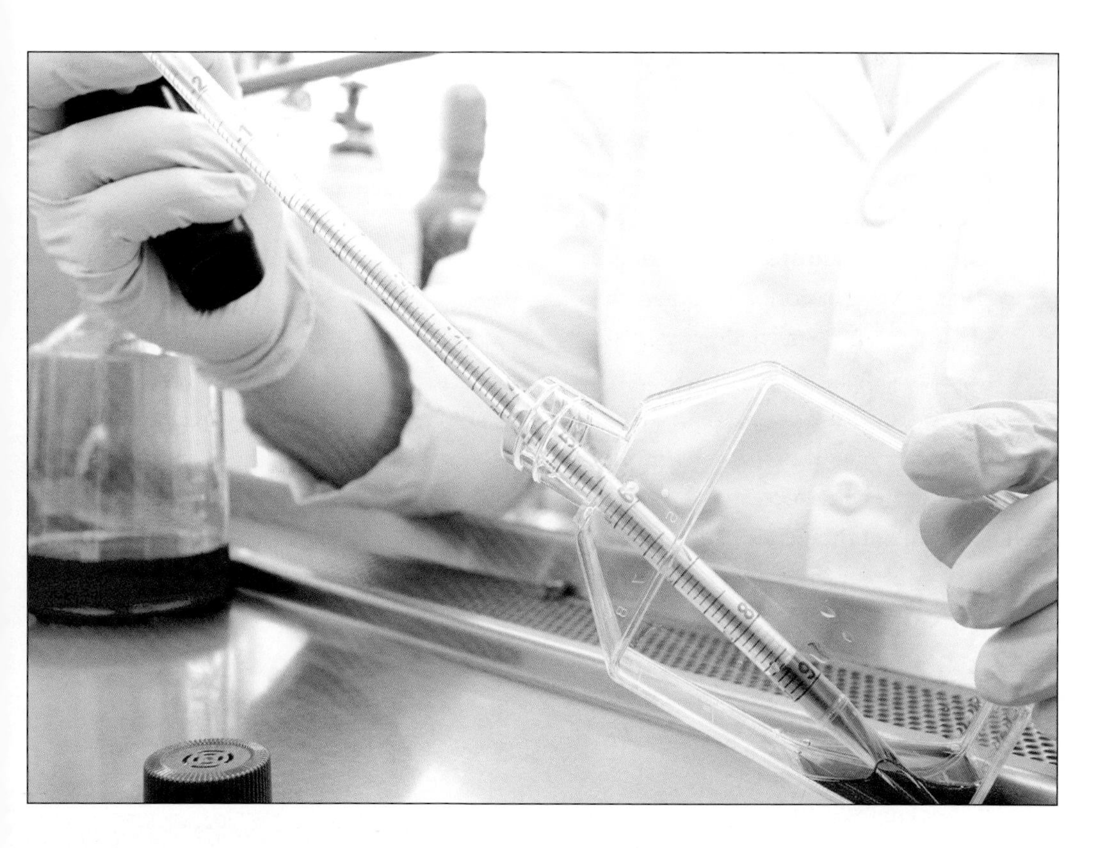

A researcher carefully prepares a stem cell specimen in a medical laboratory.

of adult stem cells is to replace specific tissue cells that are lost through natural wear and tear.

Medical researchers are investigating how to direct the development of embryonic stem cells into specific tissues. If researchers can successfully use stem cells to grow replacement body tissues and organs, many deadly diseases, such as cancer, cystic fibrosis, and diabetes, may be wiped out in the future.

The problem with research involving embryonic stem cells is that, with current techniques, the embryo must be destroyed to use its stem cells. Most embryos used by researchers are donated by couples who undergo special treatments because they cannot have children the natural way. The special treatment often involves the creation of a number of extra embryos, many of which are frozen for possible future use. When a couple does not need the frozen embryos any longer, they sometimes donate them to a laboratory that does research to help others.

Still, many people object to any research that destroys embryos. In 2001, the U.S. Congress banned all government-funded stem cell research. Only the stem cell research that was already underway could continue. Privately funded stem cell research is legal, however, and continues without the supervision of U.S. government regulations.

Another, less controversial, source of stem cells is in the umbilical cords of newborn infants. Once a baby has been born, it no longer needs the cord that linked it to its mother, so taking this blood does no one any harm. In October 2006, scientists at Newcastle University in England, working with colleagues from the National Aeronautics and Space Administration (NASA), in Houston, Texas, made an important breakthrough. For the first time ever, they succeeded in getting umbilical cord stem cells to turn into liver cells. These cells could be used in the testing of new drugs, cutting down on animal testing and possibly hazardous tests on human volunteers.

The possibilities for using stem cells in medicine are huge. They could potentially be used to provide new liver, nerve, or heart cells, for example, to replace tissues damaged by accident or disease.

Seeing Inside

Until just over a hundred years ago, there was no way for doctors to see what was going on inside the human body. Then in 1895, German physicist Wilhelm Roentgen (1845–1923) discovered **X-rays**. He began experiments with these new rays and found that they would pass right through things, including human flesh. Roentgen took an X-ray photograph of his wife's hand. The bones were clearly visible. For the first time, the inside of the body could be seen. Within a year, doctors in the United States were using X-rays to locate bone fractures and to find and remove bullets from gunshot wounds.

Since then, scientists have uncovered more ways to help doctors see inside their patients. **Ultrasound** machines send pulses of high-pitched sound waves into the body. These bounce back, or echo, from organs inside the body and are picked up by a probe that relays the echoes to a computer. From these echoes, the computer creates

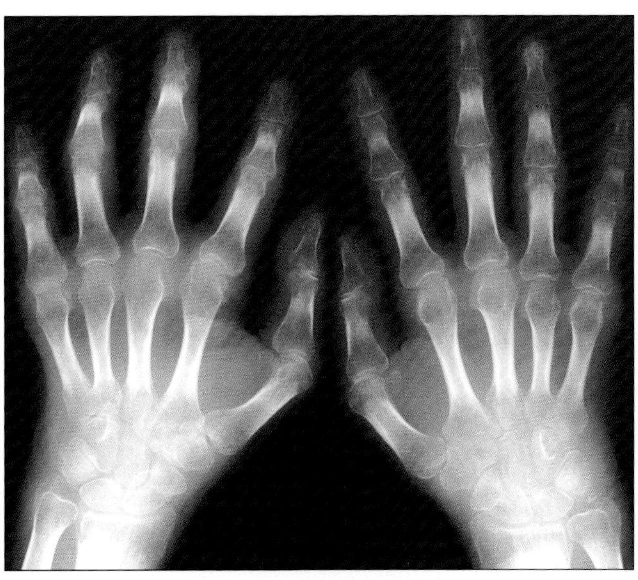

X-rays reveal the bones inside a person's hands.

a picture of the inside of the body that appears on a screen. Often ultrasound is used to check up on the progress of an unborn baby inside its mother.

In the past couple of years, ultrasound machines have been developed that can create amazing three-dimensional (3-D) pictures. These machines use sophisticated computers to build up the three-dimensional images from a number of two-dimensional (2-D) scans.

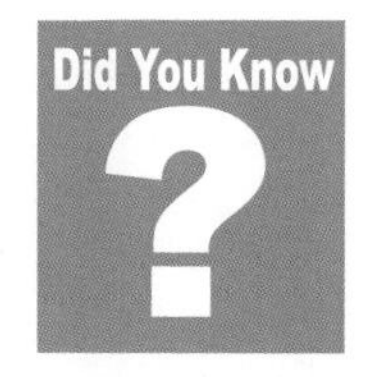

TINY BEGINNINGS
From a single stem cell (the fertilized egg) we grow into a complex organism of 100 trillion cells.

Three-dimensional ultrasound scans can show unborn babies in fascinating detail.

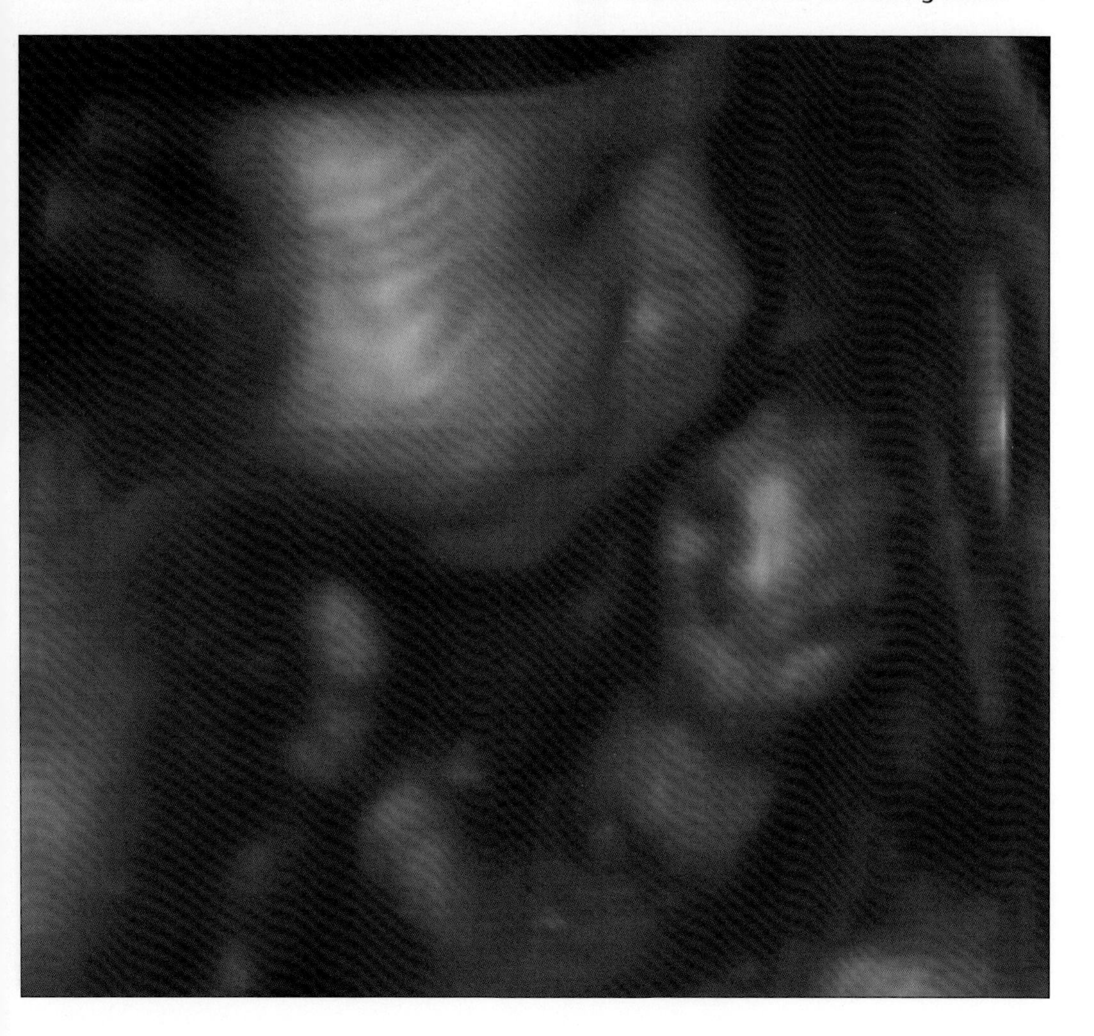

Magnetic resonance imaging (**MRI**) uses a combination of radio waves and powerful magnets to build up a remarkably detailed portrait of the inside of the human body. The pictures produced by MRI are so clear that they are often used to help diagnose sports injuries. Even tiny tears in muscles and slight damage to joints will show up in an MRI scan. Major organs such as the heart, lungs, and kidneys can also be examined for defects or problems using MRI.

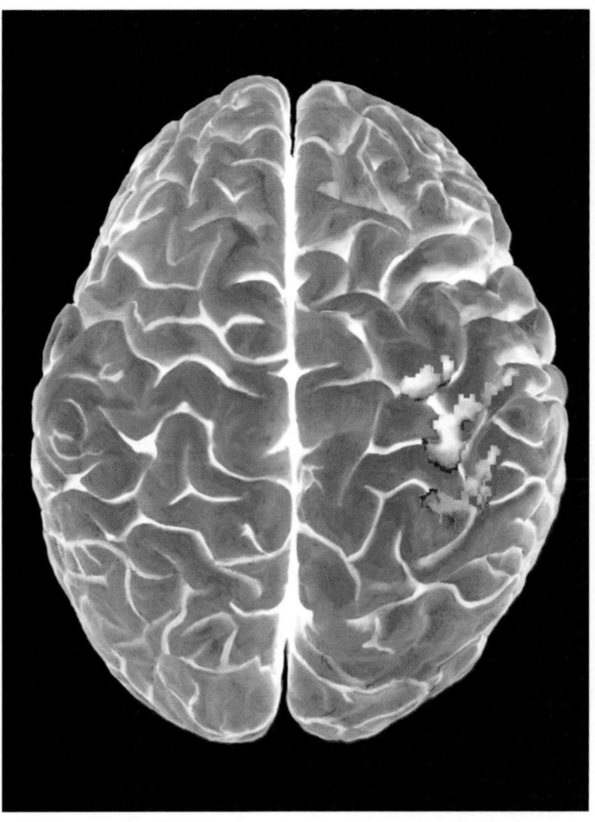

An fMRI brain scan image—the red area—shows the part of the brain that is active when the left hand is being used.

Since the 1990s, a technique called functional magnetic resonance imaging (fMRI) has been used to help our understanding of how the brain works. These images can show the different parts of the brain that become active when various tasks are performed, such as moving the fingers, looking at an object, or speaking. This is very important to brain surgeons. It can help them avoid damaging critical areas of the brain when performing surgery.

In 2006, fMRI researchers from Cambridge University in England made a remarkable discovery when they scanned the brain of a woman in a coma. The woman had been injured in a car accident and had been unconscious for more than five months. Doctors had no hope of her ever recovering. During an fMRI scan, however, the woman was asked to imagine playing tennis or walking around in her house. She didn't show any outward signs, but the scan revealed that the activity in her brain was identical to that of a healthy person. This discovery has changed the way in which comatose patients are looked after.

Micromedicine

In the 1966 science fiction movie *Fantastic Voyage*, a submarine and its crew are miniaturized and injected into someone's bloodstream. Impossible though that might be, researchers have already developed micro-robots scarcely bigger than the period at the end of this sentence. These robots can move through blood. They are equipped with an elbow, a wrist, a hand, and fingers. The researchers hope that the robots will be able to capture single cells and bacteria for analysis. They also hope to make micro-robots that might, for instance, carry a dose of drug right to the place where it will do most good. Scientists have already made a silicon microrobot half the thickness of a human hair that is powered by a heart muscle cell from a rat.

In this photograph, an ant carrying a microchip in its jaws gives an idea of just how small technology can be.

Science and Communication

Do you have a cell phone? Do you use a computer at home and at school? Do you like to watch television or listen to music on your MP3 player? None of these things would have been possible without the work of scientists.

Keeping in Touch

In one field perhaps more than any other, science has transformed our society by changing the ways we keep in touch with each other. That field is communications. Advances in telecommunications have made it possible for us to see and speak to people all over the world and to share our ideas with them almost instantaneously.

Sometimes the ideas that lead to new technology can come from unexpected places. At the beginning of the twentieth century, scientists such as Albert Einstein and Niels Bohr were thinking about what the universe was actually made of, and how it worked on the smallest levels imaginable. Einstein had the idea that light was made up of little packets of energy called **photons**. He also thought that there should be a way to direct these energy packets into a high-intensity stream of light.

Scientists working at a Hughes Aircraft Company laboratory successfully built the first **laser** beam in 1960. By 1962, a laser beam fired from Earth lit up the surface of

ALBERT EINSTEIN

Albert Einstein (1879–1955) was born in Ulm, Germany. At school, he didn't impress his teachers. He even failed the entrance exams needed to study electrical engineering. Recognizing his own ability for "abstract and mathematical thinking," he succeeded in earning a degree to teach math and physics. Unable to find a teaching job, he worked as a junior clerk in a patent office in Bern, Switzerland. He produced much of his scientific work during his spare time, including three landmark pieces that transformed our understanding of physics. Among these was his theory of relativity in which he demonstrated that Newton's laws of motion didn't apply to objects traveling at close to the speed of light and that mass and energy were linked. Einstein came to live in the United States in 1932 and made many contributions to American scientific life.

the Moon. Scientists also developed a way to use pulses of laser light fired down tubes of glass to send communications. This was no ordinary glass, but glass of incredible purity stretched into thin "hairs," called optic **fibers**. The first **fiber optic** communications line was set up between two Bell Telephone offices in Chicago, Illinois, in 1978. It was used to carry voice, data, and video signals. By 1984, a cable less than an inch (2.54 cm) in diameter was carrying 80,000 telephone conversations between cities such as Boston, Massachusetts, and Washington, D.C.—all at the same time!

The *SMART-1* space probe successfully tested the use of lasers beams to communicate with probes in deep space in 2004.

The Information Revolution

Fiber optic communications began to link cities across the United States in the 1980s. The first transatlantic fiber optic cable was installed in 1988, linking North America and France. With fiber optic cables spanning countries and oceans, a rapid and reliable communications network developed. Information in large quantities could be sent from place to place and person to person.

In 1986, the U.S. National Science Foundation set up the National Science Foundation Network (NSFNET) to connect supercomputers across the country using high-speed fiber optic cables that allowed scientists to share information. Regional networks were set up to link schools and

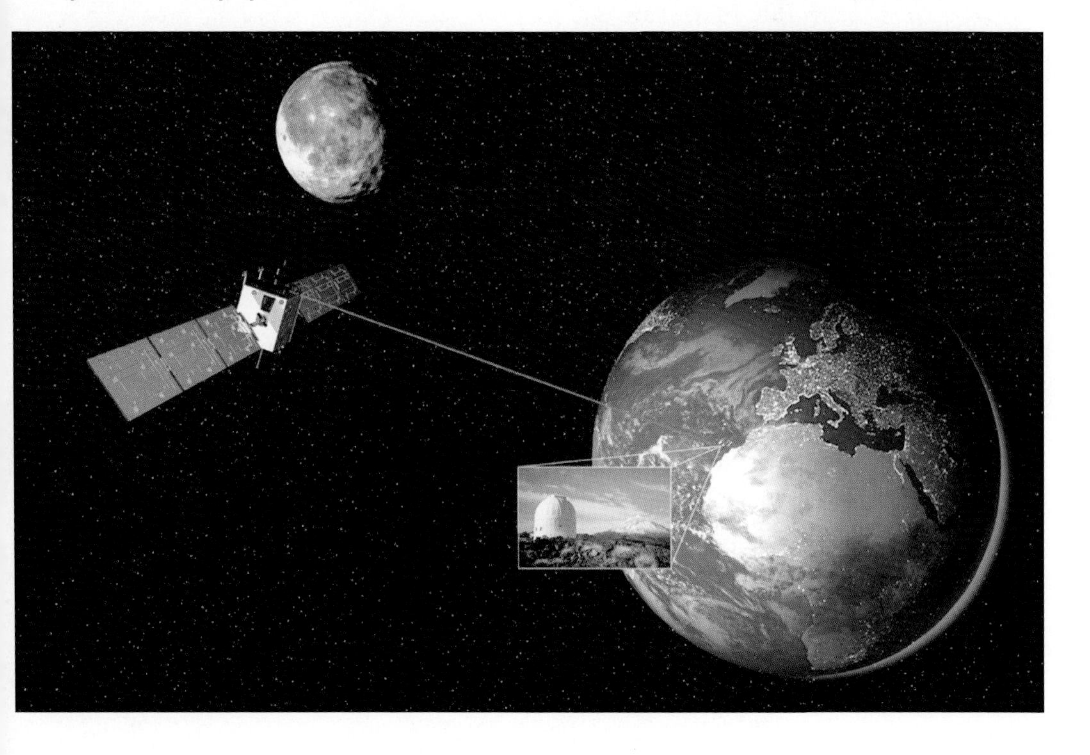

universities to NSFNET. Soon more than fifty thousand computers were linked, some from as far away as New Zealand and Brazil. They joined the network that was becoming known as the Internet.

At the same time that the network was taking shape, affordable personal computers (PCs) were becoming available. By 1984, PCs were selling at a rate of two million a year. Today, more than 175 million computers are sold each year around the world. The Internet makes use of the same network of cables that carry telephone calls, so it was inevitable that home computer users would make use of these cables.

The first **modems** were slow. These devices linked computers together using telephone lines. It would have been impossible to carry video links or to download large audio files at the slow speeds available in the late 1980s. The modems worked well enough for people living far apart to post messages to each other on electronic bulletin boards where they could discuss topics of interest.

British computer scientist Tim Berners-Lee, working at CERN (from the French name Conseil Européen pour la Recherche Nucléaire, or European Council for Nuclear Research), near Geneva, Switzerland, wanted to simplify information sharing across networks. In 1989, he proposed a project he called the World Wide Web. It was designed to allow people to combine their knowledge in a way that made it easy to follow links from one topic to another. By the summer of 1991, the World Wide Web program was available on the Internet.

Internet cafes are places where people can hook up to the Internet to find information and communicate with other people across the world.

The Web made accessing information on the Internet easy. People didn't need to know anything about the Internet's complex web of connections to get to the information they wanted. Today, we point and click from one thing to another almost without thinking about it. With the fast speeds available, users can listen to radio stations, catch up on television broadcasts, and download music and movies, all from desktop computers. Within less than a century, Einstein's packets of light have linked the world together.

Cell Phones

Cell phones are marvelous gadgets that allow us to keep in touch with family and friends at all times. A cell phone is actually a type of radio. These phones are called cell phones (the word cell is short for "cellular") because each local area is divided up into blocks of about 10 square miles (26 sq kilometers) called cells. Each cell has its own base station containing radio equipment to transmit the calls. As a phone user moves from one cell to another, the signal is passed from one base station to the next.

Because of the way they work, cell phones produce a small amount of **electromagnetic radiation**. You may have seen or heard news stories about this and the possible health risks involved. Cell phones operate at power levels too low to cause injury. FDA has also said that current scientific evidence does not suggest that any health problems are connected to cell phone use. On the other hand, the World Health Organization (WHO) advises people to be cautious about cell phone use. The organization is particularly concerned that children should keep phone use to a minimum. Although researchers continue to study the possible consequences of cell phone use, no hard and fast evidence has yet been found to link them to ill health.

A cell phone base station transmits signals from one cell phone to another.

MP3 MUSIC

MP3 is a way of compressing audio files. Without loss of quality, files compressed this way take up less space and can be transmitted more quickly. This technology was developed by Karlheinz Brandenburg and Dieter Seltzer in 1991. Today, downloading and sharing MP3 files from the Internet has become one of the most popular ways for people to listen to music. New music is available at the click of a computer mouse.

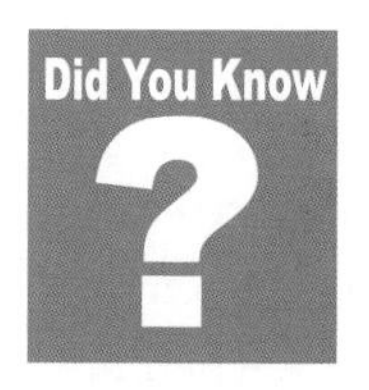

SILENT TRIBUTE

When Alexander Graham Bell, inventor of the telephone, died on August 2, 1922, there were about 14 million telephones in the United States alone. On the day of his funeral, the United States and Canada paid tribute to him by closing down their entire telephone networks for a minute's silence.

Science and Energy

In the past, it seemed we had a plentiful supply of the fossil fuels—oil, gas, and coal. Now we can see some of our fuel resources beginning to run out. Some people believe we will be out of oil around the middle of the century unless massive new reserves are found. Today we are very aware of the need to find alternative fuels to meet our energy needs.

Fuel Reserves

A hundred years ago, **prospectors** searching for oil simply looked for places where oil leaked out onto the ground. Then they started digging with little more than picks and shovels. Today, oil exploration is a science.

As part of their work, geologists—scientists who study Earth—have drawn us a picture of the way oil reserves are formed and where they might be found. Geologists use a variety of tools to map the rocks beneath the surface. The **gravimeter** is an instrument that measures minute changes in the pull of gravity. Low readings can indicate the presence of oil-bearing rocks. The **magnetometer** measures changes in Earth's magnetic field. Rocks that might hold oil have a weaker magnetic field than other types of rock. Another instrument, the **seismograph**, measures the speed of vibrations traveling through Earth. Geologists can map out different rock formations by measuring how the speed of the vibrations

America's first oil well, drilled in Titusville, Pennsylvania, in 1859. This marked the beginning of the massive U.S. petroleum industry.

changes as it moves from one to another.

Today, we are still finding new reserves of oil and gas. The problem is that these deposits are harder to get to. In 2006, the Chevron Corporation announced a major new discovery beneath the Gulf of Mexico. To get to this new deposit, however, was not easy. The prospectors first had to lower the drill through 7,000 feet (2,100 meters) of water to reach the seabed. Then they drilled another 20,000 feet (6,000 m) before they hit the oil. It is expensive to extract oil from such a depth. It can cost $100 million to set up an oil rig in waters this deep. It is a big find, but at current levels of consumption it will supply the United States for only another two years.

Another problem is that new oil reserves may be located under wilderness areas. This leads to big arguments. Politicians, oil companies, and environmentalists argue about whether or not areas such as the Arctic National Wildlife Refuge, in Alaska, should be opened for oil exploration. One side of the argument says that we can't ignore the fact that there may be two billion barrels of oil available there. The other side is that the damage done to the wilderness would be permanent—all for around three to four months' worth of the United States' oil needs.

The Trans-Alaska oil pipeline crosses the Tanana River on its way south.

What's the Alternative?

There are a number of alternatives to using fossil fuels. None of them offer perfect solutions—and some come with problems of their own.

THE RIGHT QUESTIONS

"The scientist is not a person who gives the right answers, he's one who asks the right questions."

— Claude Lévi-Strauss (b.1908), French philosopher and anthropologist

The use of **solar power**—energy from the Sun—is one possibility of an alternative fuel source. The Sun is an immense source of energy. Every day, it generates more energy than the world's entire population of six billion people could use in thirty years. How can we capture and use some of that energy?

An array of solar panels capture and harness the energy of the Sun.

One method involves using solar thermal collectors. These heat-absorbing panels use solar power to heat water. As the cost of this technology gets lower, installing these panels is becoming a popular way to save on energy bills. The panels can also be used to produce electricity by heating the water enough to produce steam to run turbines.

Photovoltaic cells turn light into electricity. They were invented at AT&T's Bell Laboratories, in New Jersey, in 1954. Originally, the cells were used to provide power for satellites and space vehicles. Now they are being used to generate electricity for homes and businesses around the world. Thousands of homes across the United States rely on solar power for either some or all of their energy needs.

Other alternatives might come from wind and waves. Wind power can be harnessed to generate electricity using wind turbines. Ocean waves and tides can also be used to run generators. The most important factor in the effectiveness of a wind farm is its location. For example, the states of North Dakota and California have good, steady winds. Wind farms take up large areas of land, but when they are built on farms and ranches, the land serves more than one purpose. Wind power has huge potential and is currently the world's fastest growing alternative energy resource. One study estimated that in the future wind farms could supply half the United States' electricity needs.

Wind farms such as this could make huge contributions to our energy needs.

Nuclear Power

Just a century ago, no one even dreamed of a resource such as **nuclear power**. In 1905, Albert Einstein said that mass could be converted into energy and vice versa. By 1918, Sir Ernest Rutherford (1871–1937) had shown that **atoms** could be split. By 1942, the world's first nuclear reactor had been built for research purposes at the University of Chicago in Batavia, Illinois.

That first reactor was built as part of the Manhattan Project. The goal of this enterprise was to produce an atomic weapon during World War II. In 1946, after the war, the Atomic Energy Commission was established. This group oversees the use of nuclear power in civilian life. The first commercial nuclear power plant in the United States went into operation in Shippingport, Pennsylvania, in 1957. Today, the United States has about a hundred nuclear power facilities. They supply about one-fifth of the country's electricity.

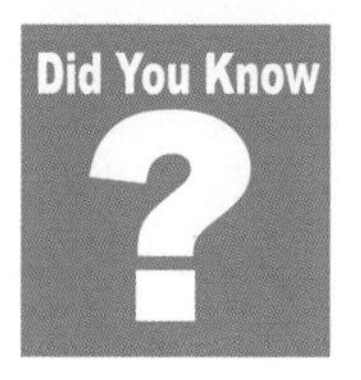

NUCLEAR POWER
There are about 440 nuclear power stations around the world, together supplying 16 percent of our energy needs.

Nuclear Waste

One of the major problems associated with nuclear power is how to deal safely with the hazardous **radioactive** waste produced. Low-level waste consists of materials such as contaminated clothing. Of greatest concern is the high-level waste such as spent fuel rods. Spent fuel generates a great deal of heat and radioactivity and has to be kept cool for decades. Radioactive waste can either be stored or disposed of. Storage means keeping it accessible for possible future use.

The Diablo Canyon nuclear power plant generates 2200 megawatts of electricity, supplying power for over two million people in north and central California.

Disposal means putting the waste safely and permanently out of reach. The U.S. Department of Energy (DOE) is planning a massive underground storage area for the country's nuclear wastes in Yucca Mountain, Nevada. However, it may be 2017 before the facility is open. Meanwhile, radioactive wastes are stored in 120 locations scattered across thirty-nine states.

Scientists cannot say with total certainty that a nuclear waste disposal site will remain secure for the thousand or more years it takes for radioactive waste to become harmless. It is very difficult to predict what changes may take place over such a long period. If radioactive wastes had been stored before Christopher Columbus set sail, they would still be harmful today. Just think how much the world has changed since then!

Nuclear Weapons

Nuclear weapons are a good illustration of human ingenuity being used for a terrible purpose. The discovery that splitting the atom would open up a new source of energy for society was an exciting one. But, as we've seen, the Manhattan Project's purpose was not to find ways to benefit society. Its goal was to turn atomic power into a weapon. When the first atomic bomb was detonated in a desert near Los Alamos, New Mexico, on July 16, 1945, it changed the world.

The incredible destructive force of a nuclear bomb is the downside of atomic energy.

The International Atomic Energy Commission has the job of monitoring the use of nuclear technology. The commission makes sure that nuclear energy is used only for peaceful purposes. It sends inspectors to countries to make sure they are not engaged in weapons-building programs. But there is little the commission can do if a country refuses to cooperate. In 1998, tensions built between India and Pakistan over nuclear experiments. At the time, India conducted weapons tests near the border between the two countries. In 2006, North Korea carried out weapons tests, too. Many governments are concerned that Iran is developing the technology needed to develop weapons.

Fusion Future?

The energy from nuclear power plants comes from splitting atoms. Nuclear **fusion** works in the opposite way. It occurs when two lighter atoms combine to form a single heavier one. The resulting atom has a smaller mass than the two lighter ones combined because some of the mass is converted into energy. This is the process that powers stars such as the Sun. Ounce for ounce, fusion produces a million times more energy than could be obtained by burning the same weight of oil. Fusion is not only a wonderful energy source, but it also produces no radioactive waste. The by-product of fusion is harmless helium gas.

Research is being carried out on fusion energy around the world, particularly in the United States, Japan, and various European countries. However, it is a very costly process. So far, more than $20 billion has been spent over forty years—and there is little sign of success. It is not likely that fusion power will be commercially available at any time within the next few decades.

Saving Energy

What can ordinary people do to help solve our energy problems? One idea is to reduce our use of energy where possible. Along with that, we could look at cutting down on waste. We use a lot of energy to heat our homes, schools, and places of work in the winter. We also use energy to cool them in summer. Electricity is used to light our towns and cities. The countless gadgets that are a part of our lives—televisions, refrigerators, dishwashers, microwaves, and computers—all use energy, too. So, too, do the factories that make all these things that we consumers are constantly demanding. The clothes we wear, the food we eat, and the homes we live in require energy to produce and maintain.

Most of this energy comes from fossil fuels, which are also used to power our cars and other vehicles. Today's cars are much more efficient than they were twenty years ago. They can travel much further on a single tank of gasoline. The downside is that there are many more cars on the road today, and we're using them more often. One promising partial solution is the use of **hybrid cars**. These cars combine gasoline-powered internal combustion engines with electrical power. The engine acts as a generator to recharge the vehicle's batteries. Once the batteries are fully charged, the car switches over to electrical power.

Hybrid cars such as this use a combination of battery power and gasoline-powered engine.

In recent years, we have watched the price of gasoline double and more. At the same time, we have come to understand more about the dangers of **pollution** and **global warming**. Many people now see that energy efficiency makes good sense! It isn't just about saving money—it's about saving the environment, too.

The Energy Star program was set up jointly by the Environmental Protection Agency (EPA) and the Department of Energy. The goal of this program is to promote the use of energy-efficient products. The Energy Star program sets standards regarding how efficiently a product such as a refrigerator or lightbulb uses energy. Products that meet the standards are awarded the Energy Star label. An Energy Star lightbulb, for example, uses two-thirds less energy than an ordinary bulb.

If we recycle as much as possible, we can save a lot of energy. It takes about one-third less energy to make paper from recycled magazines and newspapers than it does to make paper from new materials. The energy savings in making glass from recycled materials is about the same. Recycling aluminum cans is an easy way to save energy. It takes 90 percent less energy to make an aluminum can from recycled materials than it does to make one from scratch. When possible, it also saves energy to reuse some products. For example, you might wash a plastic cup and use it again instead of throwing it away.

Workers in Mumbai, India, sort through materials for recycling.

BIG PROBLEMS; SMALL SOLUTIONS?

It's easy to look at a problem and think that it is just too big to solve. But making even a small change makes a difference. Little steps together become a big step. For example, you can save energy by switching off lights when you're not in the room. Turn off television sets and computers; don't leave them on standby. Don't throw away cans and bottles—recycle them. The message is: reduce, reuse, recycle.

Planet Science

At the beginning of the twenty-first century, we are facing some big problems. In fact, we might say that they are planet-sized problems. We've just looked at the problem of energy. There is more to energy, however, than just making sure we have enough. There are the consequences of producing it, too.

Removing fossil fuels from the ground and transporting them to where they are to be used has an impact on the environment. Waste materials from deep coal mines are piled in huge dumps on land or disposed of at sea. Leaks from oil wells in shallow coastal waters pollute beaches and kill sea life. Oil tanker and pipeline breaks can cause serious environmental harm. For example, the 1989 *Exxon Valdez* oil spill caused so much damage in Alaska that the U.S. Congress passed the Oil Pollution Act of 1990. This act demands that shipping companies use double-hulled oil tankers to carry oil. This practice, it is believed, reduces the chances of oil being spilled if the vessel runs aground. Shipping companies have been given until 2015 to upgrade their fleets.

THE EXXON VALDEZ

When it ran aground in Prince William Sound, Alaska, in March 1989, the *Exxon Valdez* dumped 11 million gallons (41.8 million liters) of crude oil and contaminated about 1,300 miles (2,080 km) of the Alaskan coastline. About one-quarter of a million seabirds, nearly 3,000 sea otters, 300 harbor seals, and at least twenty killer whales died as a result of the spill.

Since the accident, the *Exxon Valdez* has been repaired and renamed the *Sea River Mediterranean*. It is now working in the Atlantic, but it is banned from ever returning to Alaska.

Air Pollution

Air pollution is hazardous to Earth's people, its wildlife, and its ecosystems. The major sources of air pollution are cars and other vehicles, factories, and fossil fuel-burning power plants.

Acid precipitation is caused when chemicals are released into the air from sources such as vehicle exhausts and coal-burning power plants. These chemicals dissolve in water in the atmosphere to form nitric and sulfuric acids. The additional

A week after the *Exxon Valdez* disaster, teams of workers attempt to clean up the spill.

chemicals make rain, snow, and even fog more acidic than normal. Air currents may carry the acid water a long way from the actual source of the pollution. When it falls, it causes damage to crops, trees, lakes, and buildings. Some lakes in the northeastern United States are so acidic that fish can no longer live there.

Acid precipitation can be reduced. Devices called scrubbers are attached to the smokestacks of power plants. These work to remove sulfur dioxide. It is this chemical, produced when coal is burned, that forms sulfuric acid when it dissolves in water. Another way is for power plants not to burn coal at all. Natural gas power plants create much less acid precipitation than other plants. Of course, nuclear power creates none at all, but, as we have seen, that isn't a perfect solution either.

Catalytic converters on cars work like scrubbers to reduce emissions. In the United States, these devices have been a legal requirement for more than twenty years.

Climate Change

Today, many people are very concerned about climate change. Earth's atmosphere naturally contains gases, called **greenhouse gases**, such as water vapor, methane, and carbon dioxide. Energy from the Sun warms the surface of Earth, and some is reflected back into space. Greenhouse gases trap some of this reflected heat and keep Earth warmer than it would be otherwise.

Whenever fossil fuels are burned, carbon dioxide is formed. About 6.6 billion tons (6 billion tonnes) of this greenhouse gas are released into the atmosphere every year as a result of human activities. Natural processes, such as the absorption of carbon dioxide by trees and other plants, removes about half of this, but the rest remains in the atmosphere for centuries. This means that

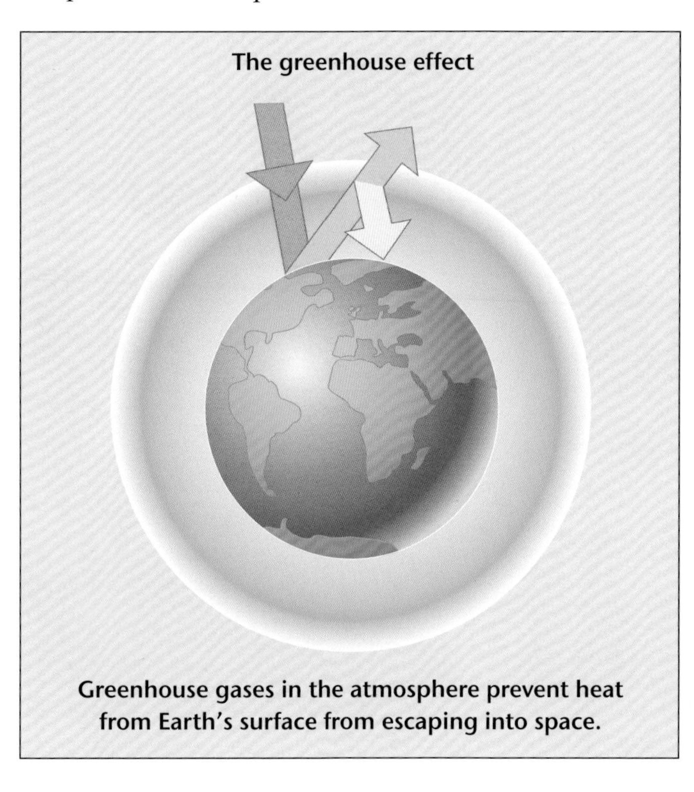

The greenhouse effect

Greenhouse gases in the atmosphere prevent heat from Earth's surface from escaping into space.

over the last two hundred years or so since we started burning fossil fuels, carbon dioxide levels have been increasing.

Many scientists believe that these rising carbon dioxide levels are causing Earth's climate to become warmer. This effect is called global warming. Across the world, we see mountain glaciers shrinking and polar ice caps melting. As polar ice melts, sea levels around the world will rise. In addition, the ocean water will expand as it gets warmer. The result could be devastating floods in coastal areas. No one really knows what the long-term weather changes will be. Some areas may see hurricanes and increased rainfall, while others may be stricken by drought.

Plants and animals are threatened with **extinction** as rising temperatures change conditions in their habitats. Some animals are shifting their ranges northward or upward on mountain slopes where conditions are cooler. For many animals, there is nowhere to go.

The Arctic walrus is threatened by changes to its habitat as global warming causes the polar ice cap to melt.

Just how much responsibility we take for this warming effect is disputed. Some people believe that what we are seeing is a result of natural changes in Earth's climate. Science still has much work to do in understanding all the factors involved, such as changing levels in the output of energy from the Sun. The vast majority of the world's climate scientists believe that most of the global warming is caused by human activity.

CERTAINTIES AND LIKELIHOODS

"Science doesn't give you certainties. Science gives you likelihood, but we think that it's likely that climate warming of the last few decades isn't due to the usual causes—changes in the Earth's orbit, changes in the Sun, volcanoes, but it's due primarily to humans adding greenhouse gases to the atmosphere."

— Richard Somerville, meteorologist and climate change expert, Scripps Institution of Oceanography

Saving the Future

In 1997, many countries signed onto the Kyoto Treaty, an international agreement addressing global warming and how to deal with it. One of the treaty's aims is to reduce greenhouse gas emissions. Some countries, such as the United States and Australia, did not agree to sign the treaty, fearing it would be damaging to their economies.

Supporters of the Kyoto Treaty demonstrate in Tokyo, Japan.

If we're going to cut carbon emissions, the most important thing to do is to save energy. More and more companies are realizing that being environmentally friendly can help them cut their costs, too. By 2003, the DuPont pharmaceutical company had cut its greenhouse gas emissions by 72 percent. The company also saved billions of dollars in energy costs by using methane gas from landfill sites in its industrial boilers rather than natural gas. In another example, the Boeing aircraft company upgraded the lighting in its facilities. This reduced the company's electric light bill by 90 percent.

Natural Disaster

In 2000, more than 250 million people around the world were affected in some way by natural disasters. Processes at work in and around Earth can cause powerful natural events. These events include earthquakes, volcanic eruptions, tornadoes, and hurricanes. We can't do anything to prevent these events. There are some things we can do to reduce the risk they pose, however. For example, we can design buildings so they have a better chance of surviving a natural disaster. We can also avoid building homes where these events are more likely to occur.

Space exploration is not NASA's only task. Part of NASA's mission is to understand and protect our home planet, Earth. For example, observations made by NASA's Earth-orbiting **satellites** can help agencies like the Federal Emergency Management Agency (FEMA) track the progress of tropical hurricanes and forest fires. This will help these other groups decide the best place to send assistance. Already, scientists all over the world use NASA satellite photos to study the causes and effects of natural disasters.

Many homes in New Orleans, Louisiana, destroyed by Hurricane Katrina in August 2005 were not built to withstand such high winds.

Tsunami Early Warning

In 2004, communities around the coasts of South Asia were devastated by a massive **tsunami**, or giant wave. The cause of the tsunami was an earthquake under the ocean. The earthquake lifted up the seabed, displaced a large amount of water, and set the wave into motion. In two hours, the tsunami rushed across a thousand miles (1,600 km) of ocean. Thousands of people lost their lives when it struck. Seismologists (earthquake scientists) recorded the earthquake but couldn't alert people in the region quickly enough to warn them of the approaching disaster.

Today, a tsunami early warning system is being put in place in the Indian Ocean. (A similar system is already in place for the Pacific Ocean.) Ocean-floor sensors detect earthquakes and send signals to buoys floating on the ocean surface. These, in turn, signal satellites orbiting overhead. Scientists monitoring the satellites will know almost at once if there is a tsunami danger. They can determine where, and how hard, it will hit, and send that information as quickly as

This photograph gives some idea of the damage caused by the tsunami that hit southern Asia in 2004.

possible to the endangered areas. This warning will give people as much time as possible to evacuate the area.

Countries that border the Indian and Pacific Oceans are now putting up warning signs to warn people that they are in a tsunami danger area.

Making Choices

Scientists collect evidence by doing experiments. This research can help them explain how and why things happen the way they do. Scientists try to use these explanations to predict what might happen in the future. But there are never any absolute certainties in science.

We can learn a great deal from science. One thing we won't learn, however, is how to make choices. Science won't do that for you. It can give you an idea of how likely something is to happen, but it won't make decisions for you. For example, science might tell you that eating junk food is bad for your health, but it won't tell you to stop eating it.

SCIENCE AND THE FUTURE

"It is science alone that can solve the problems of hunger and poverty, of insanitation and illiteracy, of superstition and deadening custom and tradition, of vast resources running to waste, of a rich country inhabited by starving people . . . Who indeed could afford to ignore science today? At every turn we have to seek its aid. . . . The future belongs to science and those who make friends with science."

— Jawaharlal Nehru (1889–1964), Indian statesman

Where will science lead in the future? Many people have their own ideas about what they'd like science to do. Doing something about energy shortages, dealing with climate change, and finding cures for fatal illnesses, such as cancer, probably come close to the top of many people's lists. But science can't do all that alone. We might not all be able to become doctors, but we can all turn lights off and recycle our soda cans. The problems we face aren't just science's problems, although science may be our best bet for finding answers. They are society's problems, and society means us, you and me, all of us, including the scientists, because science is part of society, too.

GLOSSARY

acid precipitation rain and other forms of precipitation that have been made acidic by the presence of chemicals from burning coal, vehicle exhausts, and other sources

antibiotics a substance produced by or obtained from certain bacteria or fungi that can be used as a defense against infection by disease-causing bacteria

atom tiny particles from which all materials are made. The smallest unit of matter that can take part in a chemical reaction.

catalytic converter a device fitted to the exhaust of a vehicle to cut down on the emission of harmful chemicals

cell the basic unit of life. Cells can exist as independent life-forms or form tissues in more complicated life-forms, such as the cells that form muscle tissue in animals.

climate change change in the climate of an area, or of the world, over a long period of time. Many people believe the world is currently undergoing climate change as a result of global warming.

diagnosis identifying a disease from its signs and symptoms

DNA (deoxyribonucleic acid) the genetic material of living things. DNA carries instructions for constructing, maintaining, and reproducing living cells.

electromagnetic radiation a type of energy that travels in the form of waves. Radio, microwaves, light, and X-rays are all forms of electromagnetic radiation.

embryo the earliest stage of development of an organism after the egg has been fertilized and has begun to divide, but before any organs form

experiment a practical test designed to gather information or to try out a hypothesis

extinction the complete die-off of all the members of a particular type of plant or animal in all locations

fiber a long continuous thread of a material

fiber optics thin fibers of glass or plastic through which light is transmitted

fossil fuels fuels produced by the action of heat and pressure in Earth's crust on the remains of plants and animals that lived millions of years ago. Fossil fuels include coal, petroleum, and natural gas.

fusion process by which two small atoms are made to combine, or fuse, together to form a larger atom with the release of a great deal of energy

gene a section of DNA that contains the information needed for making a single protein. Genes are the basic units of heredity.

gene therapy a medical technique for treating inherited illnesses by repairing or replacing damaged DNA

genetics the study of genes and how they work

global warming an increase of the overall temperature of the world, believed to be caused by the greenhouse effect

gravimeter a device for measuring variations in Earth's gravity

gravity the force of attraction between objects

greenhouse gas a gas in the atmosphere that absorbs heat radiated from Earth's surface that would otherwise escape into space. Carbon dioxide, methane, and water vapor are all greenhouse gases.

heredity the transfer of characteristics from parent to offspring

hybrid car a car that can run on either gasoline or electricity

hypothesis an idea, or an explanation as to why certain things have happened

immune system a body system that helps repair injury and fight disease

immunization giving protection against diseases, usually by vaccination

laboratory a place where scientific experiments are carried out

laser a device that produces a very intense, narrowly focused beam of light

law (science) an explanation or a statement that appears to be true in all cases

magnetometer a device for measuring the strength of a magnetic field

modem a device for sending data from one computer to another

MRI a device that uses powerful magnets to build up detailed three-dimensional images of the inside of the human body

nuclear power a power supply that makes use of the energy released when atoms of uranium are split

photons the smallest packages of energy in which light and other forms of electromagnetic energy travel

photovoltaic cell a device that converts light energy into electrical energy

pollution the dirtying of the land, oceans, and air

prediction a statement that a certain thing will happen

proboscis the long tongue of an insect

prospector someone who goes searching for valuable minerals, such as gold or oil

pus the thick, yellow liquid that forms in an infected wound

radioactive the ability to give off high-energy rays or particles

satellite a spacecraft in orbit around Earth that collects information or relays communication

seismograph a device that measures the speed of vibrations traveling through Earth's crust

solar power energy from the Sun

stem cells unspecialized cells that have the potential to develop into many different cell types

technology the practical application of scientific knowledge

tissue culture cells from a plant or animal that have been removed from the organism and grown under controlled conditions in a laboratory for experimentation

tsunami a powerful ocean wave caused by an earthquake or volcano. Tsunamis can travel at speeds up to 500 miles (800 km) per hour and reach as much as 100 feet (30 m) in height when they hit land.

ultrasound sound that is too high-pitched to be heard by human ears. Ultrasound waves travel through the body and can be used to "see" organs inside.

vaccine a disease-causing organism that has been made harmless before being introduced into the body to give immunity against disease

virus a disease-causing agent that can get inside a cell and cause it to make copies of the virus

X-rays a form of electromagnetic radiation that passes through many solids and makes it possible to see inside them

Books

Brinckerhoff, Richard R.
One-minute Readings: Issues in Science, Technology and Society.
Addison Wesley Publishing Co, 1992.

Discovery Channel School Science.
Genetics.
Gareth Stevens, 2003.

Michels, Dia, and Nathan Levy.
101 Things Everyone Should Know About Science.
Platypus Media, 2005.

Morris, Neil
Global Warming
What If We Do Nothing (series)
World Almanac® Library, 2007

Pringle, Laurence.
Global Warming: The Threat of Earth's Changing Climate.
SeaStar, 2003.

Sherman, Josepha.
The History of the Internet.
Franklin Watts, 2003.

Snedden, Robert.
Energy Alternatives.
Heinemann Library, 2006.

Swanson, Diane.
Nibbling on Einsteins's Brain: The Good, the Bad and the Bogus in Science.
Annick Press, 2001.

Wheeler, Benjamin, et al.
It's All Connected: A Comprehensive Guide to Global Issues and Sustainable Solutions.
Facing the Future, 2005.

Web Sites

CNN International
edition.cnn.com/TECH/space
Science in the news from CNN.

Energy Information Administration
www.eia.doe.gov/kids/energyfacts/index.html
Information on all forms of energy, including fossil fuels, nuclear power, and alternative energy sources.

Eurekalert!
www.eurekalert.org/kidsnews/
Science updates specially for kids from the American Association for the Advancement of Science.

Internet History
www.netvalley.com/intval1.html
The history of the Internet and the World Wide Web.

Internet Public Library
www.ipl.org/div/subject/browse/hea00.00.00/
Links to a number of topics on science and medicine from the Internet Public Library.

Scientific American
www.sciam.com
Scientific American magazine's website—a great place to find out about current goings on in science.

Union of Concerned Scientists
www.ucsusa.org/global_warming/
The causes and consequences of global warming from the Union of Concerned Scientists website.

Publisher's note to educators and parents: Our editors have carefully reviewed these Web sites to ensure that they are suitable for children. Many Web sites change frequently, however, and we cannot guarantee that a site's future contents will continue to meet our high standards of quality and educational value. Be advised that children should be closely supervised whenever they access the Internet.

INDEX